Dedicated to

*Mary Treat (1830–1923), who taught children and adults alike
how to study and respect nature. May those lessons live on in this book
and inspire a new generation of ecologists.*

Botanist, entomologist, naturalist, 19th century flora and fauna hunter, nature writer, teacher, pioneer, trailblazer. Mary Treat (1830-1923) was all of these, yet she is not as well-known as male scientists of the 19th century. Her study of carnivorous (insect-eating) plants overlapped with the work of Charles Darwin's studies of evolutionary plant and insect biology. Mary Treat, in one case, observed the opposite of results Darwin reported and she confidently noted his error. Darwin acknowledged his mistake and thanked her for setting the scientific record straight!

ISBN 978-1-365-24576-3

Inside illustrations by Laura Bethmann, many of them inspired
by drawings in books and articles by Mary Treat
Cover illustration by Emily Farrell
Layout by Clarissa Toney

Visit www.MaryTreat.com

Mary HAD A LITTLE ZOO

All About Pioneer Scientist Mary Treat and
Her Ants, Plants, and Spiders

by
Deborah Boerner Ein

Mary had a little zoo
Inside and out back,

Growing up to be a scientist,
She kept the zoo on track.

Her favorites were the spiders and ants
Bees, birds, and wasps, too.

She chased away the cats and dogs,
Kept safe her new bug crew.

Within a circle of bushes
Mary sat beneath a tree.

From there she watched and listened
In her backyard menagerie.

*Hey kids, Mary invites you to add whatever
awesome creatures you would like
inside and outside her circle of hedges!*

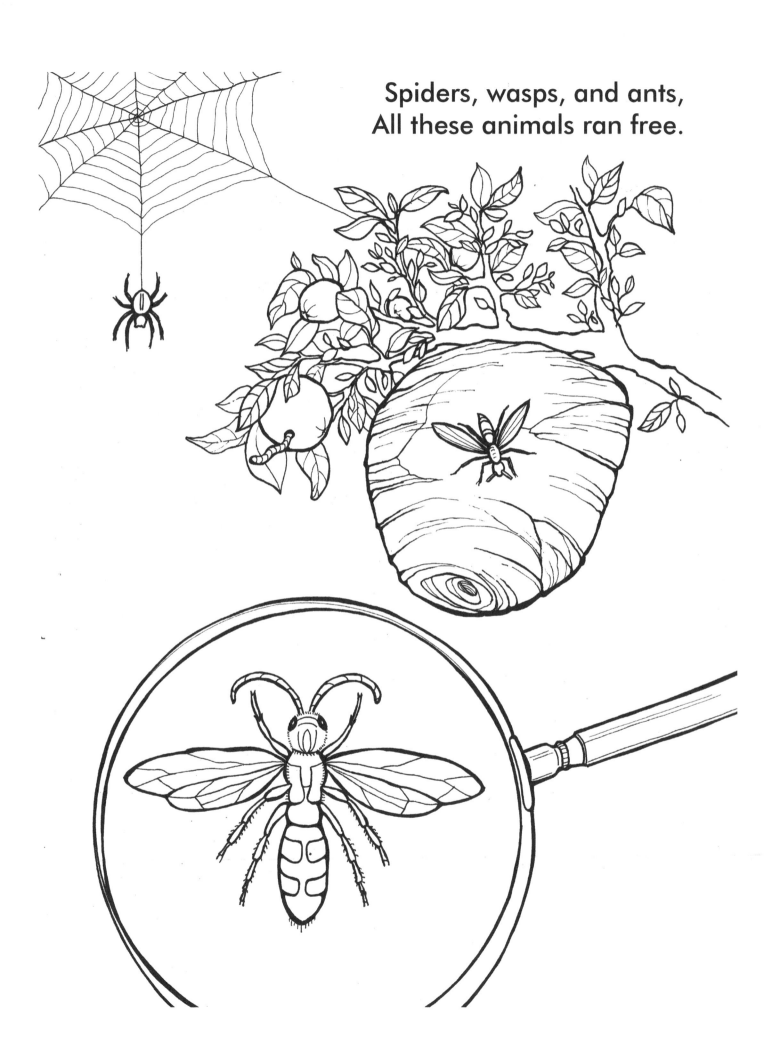

Spiders, wasps, and ants,
All these animals ran free.

And everything that Mary did,
She did for all these three.

They lived in spaces mostly hidden
In their tunnels underground,

But Mary was so very patient,
She would wait around.

Crumbs of bread and sugar grains,
Mary fed her spiders and ants.

Soon they came right up to her

And let her watch them dance!

Some she placed inside glass jars,
With dirt to build their homes.

These she brought into her house,
To watch within their domes.

Spiders crawled on leaves and sticks,
Ants tunneled along the glass.

Mary watched their comings and goings,
Oh, she had quite a blast!

A magnifier, a microscope,
Did help Mary see,

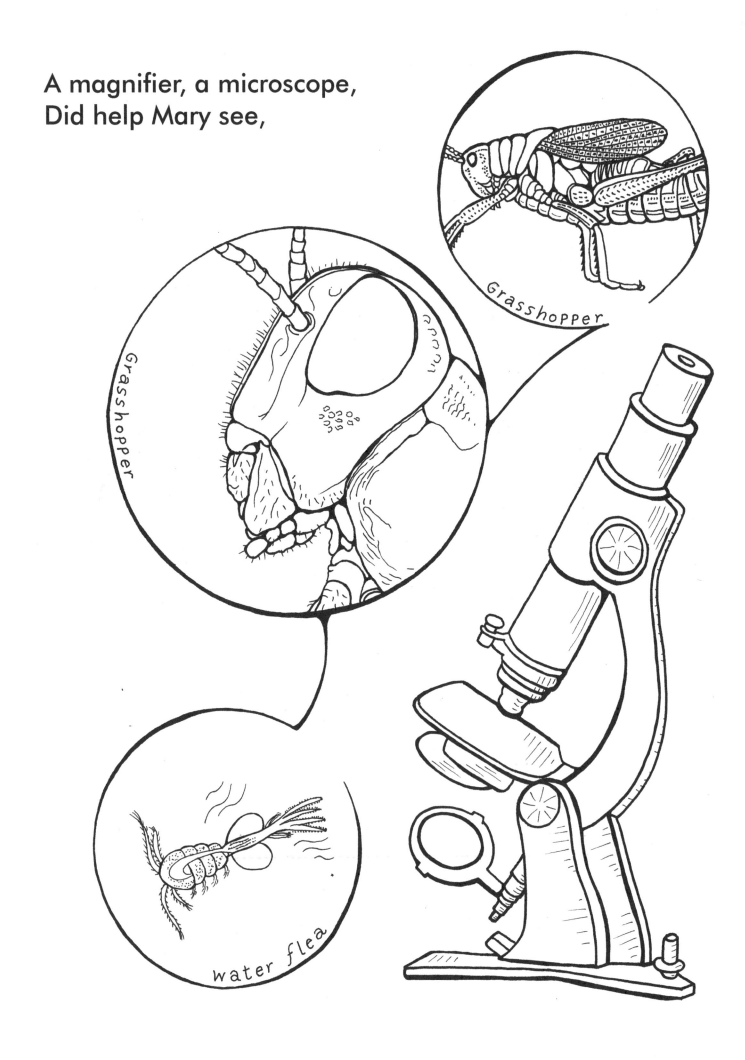

Grasshopper

Grasshopper

water flea

A miniature world of spidery legs,
Wasp and butterfly wings, with glee!

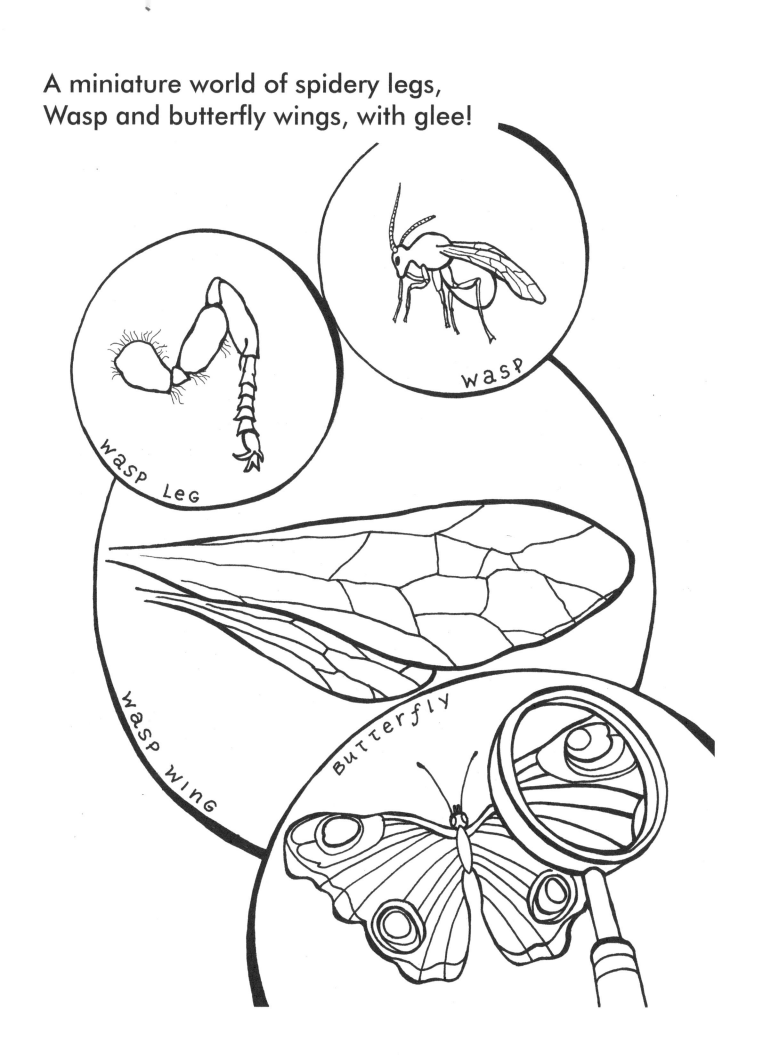

wasp

wasp leg

wasp wing

Butterfly

Mary made some cool discoveries,
She studied plants besides.

She watched a leafy plant fold up
And gobble some great big flies.

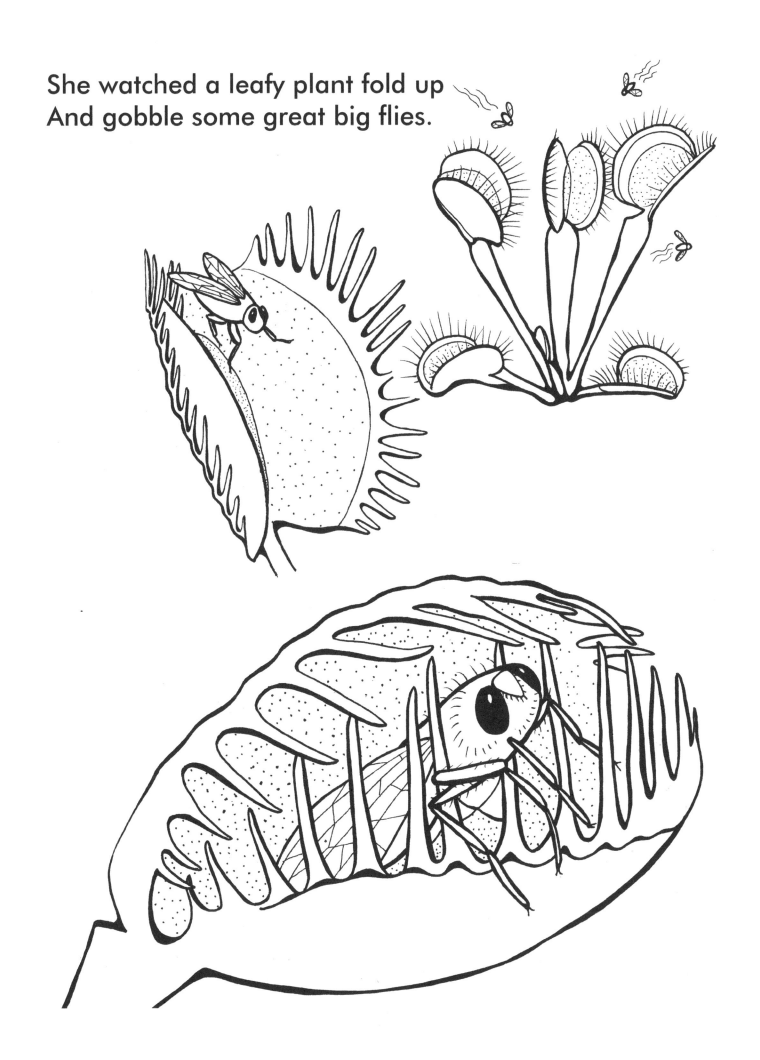

Over land and sea, in mail sent to and fro,
Mary told the other scientists all that she did know.

Charles Darwin

She wrote some books and papers, too,
For people like you and me.

Her name was Mary Adelia Treat,
She lived so long ago.

A wasp, a flower, and three ants
Are named for her, and so…

Nature holds so many secrets still
For you to find, now go!

Outside is where you'll find them
Just as Mary Treat did so.

On these pages, draw anything you like!
How about your favorite insect? Or your favorite tree?

*How about drawing a bird you see in your yard or
on a hike with family or friends?*

*Scan your drawing and ask Mom or Dad or your teacher
to visit **MaryTreat.com** for how to send your scan to
this book's author! She will post as many as possible.*

Mary Treat wrote to many famous scientists. She wrote articles in magazines and books, and she wrote stories for kids like you, too. Would you like to read some of them? Ask Mom or Dad to find a couple of those listed here! And as you get older, you can look up and read the other articles and books written by this early American scientist.

Mary Treat's book and articles for young readers:

- *Through a Microscope: Something of the Science with Many Curious Observations Indoor and Out and Directions for a Home-made Microscope* (with Samuel Wells and Frederick Sargent) (Chicago: Interstate Publishing, 1886)
- "The Water Bear," *St. Nicolas Magazine* 2 (March 1875): 274–75
- "The Cyclops," *St. Nicolas Magazine* 2 (Sept 1875): 686–87
- "The Floscule," *St. Nicolas Magazine* 3 (March 1876): 300–301
- "The Microscopic Brick Maker," *St. Nicolas Magazine* 3 (April 1876): 374–75
- "Florida Fishers," *St. Nicolas Magazine* 4 (May 1877): 490–91
- "Some Fishing Birds of Florida," *St. Nicolas Magazine* 5 (Feb 1878): 282–83
- "The Microscope and What I Saw Through It," *St. Nicolas Magazine* 6 (Dec 1878): 116–17
- "Our Cuckoos and Cowbirds," *Youth's Companion* (1887)

Add yourself to this picture and send it to your friends!